PENGUIN BOOKS — GREAT IDEAS

The Art of War

Sun-tzu

c. 551–496 BC

Sun-tzu

The Art of War

TRANSLATED BY
JOHN MINFORD

PENGUIN BOOKS — GREAT IDEAS

PENGUIN BOOKS

Published by the Penguin Group
Penguin Books Ltd, 80 Strand, London WC2R 0RL, England
Penguin Group (USA) Inc., 375 Hudson Street, New York, New York 10014, USA
Penguin Group (Canada), 10 Alcorn Avenue, Toronto, Ontario, Canada M4V 3B2
(a division of Pearson Penguin Canada Inc.)
Penguin Ireland, 25 St Stephen's Green, Dublin 2, Ireland
(a division of Penguin Books Ltd)
Penguin Group (Australia), 250 Camberwell Road,
Camberwell, Victoria 3124, Australia (a division of Pearson Australia Group Pty Ltd)
Penguin Books India Pvt Ltd, 11 Community Centre,
Panchsheel Park, New Delhi – 110 017, India
Penguin Group (NZ), cnr Airborne and Rosedale Roads, Albany,
Auckland 1310, New Zealand (a division of Pearson New Zealand Ltd)
Penguin Books (South Africa) (Pty) Ltd, 24 Sturdee Avenue,
Rosebank 2196, South Africa

Penguin Books Ltd, Registered Offices: 80 Strand, London WC2R 0RL, England

www.penguin.com

014

A portion of this work first appeared in *The New England Review*
First published in the United States of America by Viking Penguin 2002
Published in Penguin Books 2005

Translation copyright © John Minford, 2002
All rights reserved

Set by Rowland Phototypesetting Ltd, Bury St Edmunds, Suffolk
Printed in England by Clays Ltd, St Ives plc

ISBN-13: 978–0–141–02381–6

www.greenpenguin.co.uk

ALWAYS LEARNING   **PEARSON**

Contents

Making of Plans

Master Sun said:

> War is
> A grave affair of state;
> It is a place
> Of life and death,
> A road
> To survival and extinction,
> A matter
> To be pondered carefully.

There are Five Fundamentals
> For this deliberation,
> For the making of comparisons
> And the assessing of conditions:
> The Way,
> Heaven,
> Earth,
> Command,
> Discipline.

The Way
> Causes men
> To be of one mind
> With their rulers,

To live or die with them,
And never to waver.

Heaven is
Ying and Yang,
Cold and hot,
The cycle of seasons.

Earth is
Height and depth,
Distance and proximity,
Ease and danger,
Open and confined ground,
Life and death.

Command is
Wisdom,
Integrity,
Compassion,
Courage,
Severity.

Discipline is
Organization,
Chain of command,
Control of expenditure.

Every commander is aware
Of these
Five Fundamentals.
He who grasps them

Wins;
He who fails to grasp them
Loses.

For this deliberation,
 For the making of comparisons,
 And the assessing of conditions,
 Discover:

 Which ruler
 Has the Way?

 Which general
 Has the ability?

 Which side has
 Heaven and Earth?

 On which side
 Is discipline
 More effective?

 Which army
 Is the stronger?

 Whose officers and men
 Are better trained?

 In which army
 Are rewards and punishments
 Clearest?

From these
    Can be known
    Victory and defeat.

Heed my plan,
    Employ me,
    And victory is surely yours;
    I will stay.

Do not heed my plan,
    And even if you did employ me,
    You would surely be defeated;
    I will depart.

Settle on the best plan,
    Exploit the dynamic within,
    Develop it without,

Follow the advantage,
    And master opportunity:
    This is the dynamic.

The Way of War is
    A Way of Deception.

    When able,
    Feign inability;

    When deploying troops,
    Appear not to be.

When near,
Appear far;

When far,
Appear near.

Lure with bait;

Strike with chaos.

If the enemy is full,
Be prepared.
If strong,
Avoid him.

If he is angry,
Disconcert him.

If he is weak,
Stir him to pride.

If he is relaxed,
Harry him;

If his men are harmonious,
Split them.

Attack
Where he is
Unprepared;
Appear

Where you are
Unexpected.

This is
 Victory in warfare;
 It cannot be
 Divulged
 In advance.

Victory belongs to the side
 That scores most
 In the temple calculations
 Before battle.
Defeat belongs to the side
 That scores least
 In the temple calculations
 Before battle.
Most spells victory;
 Least spells defeat;
 None, surer defeat.
I see it in this way,
 And the outcome is apparent.

Waging of War

Master Sun said:

In War,
>   For an army of
>   One thousand
>   Four-horse swift chariots,
>   One thousand
>   Hide-armoured wagons,
>   For one hundred thousand
>   Mail-clad soldiers,
>   With provisions for
>   Four hundred miles;

Allowing for
>   Expenses at home and at the front,
>   Dealings with envoys and advisers;
>   Glue and lacquer,
>   Repairs to chariots and armour;
>   The daily cost of all this
>   Will exceed
>   One thousand taels of silver.

In War,
>   Victory should be
>   Swift.
>   If victory is slow,

Men tire,
Morale sags.
Sieges
Exhaust strength;
Protracted campaigns
Strain the public treasury.

If men are tired,
Morale low,
Strength exhausted,
Treasure spent;
Then the feudal lords
Will exploit the disarray
And attack.
This even the wisest
Will be powerless
To mend.

I have heard that in war
Haste can be
Folly
But have never seen
Delay that was
Wise.

No nation has ever benefited
From a protracted war.

Without a full understanding of
The harm
Caused by war,

It is impossible to understand
The most profitable way
Of conducting it.

The Skilful Warrior
    Never conscripts troops
    A second time;
    Never transports provisions
    A third.

    He brings equipment from home
    But forages off the enemy.
    And so his men
    Have plenty to eat.

Supplying an army
    At a distance
    Drains the public coffers
    And impoverishes
    The common people.

Where an army is close at hand,
    Prices rise;
    When prices rise,
    The common people
    Spend all they have;
    When they spend all,
    They feel the pinch of
    Taxes and levies.

Strength is depleted
    On the battlefield;
    Families at home
    Are destitute.

The common people
    Lose seven-tenths
    Of their wealth.
    Six-tenths of the public coffers
    Are spent
    On broken chariots,
    Worn-out horses,
    Armour and helmets,
    Crossbows and arrows,
    Spears and bucklers,
    Lances and shields,
    Draft animals,
    Heavy wagons.

So a wise general
    Feeds his army
    Off the enemy.
    One peck
    Of enemy provisions
    Is worth twenty
    Carried from home;
    One picul
    Of enemy fodder
    Is worth twenty
    Carried from home.

The killing of an enemy
    Stems from
    Wrath;
The fighting for booty
    Stems from
    A desire for reward.

In chariot fighting,
    When more than ten
    Enemy chariots are captured,
    The man to take the first
    Should be rewarded.
Change the enemy's
    Chariot flags and standards;
    Mingle their chariots
    With ours.

Treat prisoners of war kindly,
    And care for them.
    Use victory over the enemy
    To enhance your own strength.

In War,
    Prize victory,
    Not a protracted campaign.

The wise general
    Is a Lord of Destiny;
    He holds the nation's
    Peace or peril
    In his hands.

Strategic Offensive

Master Sun said:

In War,
    Better take
    A state
    Intact
    Than destroy it.

Better take
    An army,
    A regiment,
    A detachment,
    A company,
    Intact
    Than destroy them.

Ultimate excellence lies
    Not in winning
    Every battle
    But in defeating the enemy
    Without ever fighting.
    The highest form of warfare
    Is to attack
    Strategy itself;

The next,
　　To attack
　　Alliances;

The next,
　　To attack
　　Armies;

The lowest form of war is
　　To attack
　　Cities.
　　Siege warfare
　　Is a last resort.

In a siege,
　　Three months are needed
　　To assemble
　　Protective shields,
　　Armoured wagons,
　　And sundry
　　Siege weapons and equipment;
　　Another three months
　　To pile
　　Earthen ramps.

The general who cannot
　　Master his anger
　　Orders his troops out
　　Like ants,
　　Sending one in three
　　To their deaths,

Without taking the city.
This is the calamity
Of siege warfare.

The Skilful Strategist
Defeats the enemy
Without doing battle,
Captures the city
Without laying siege,
Overthrows the enemy state
Without protracted war.

He strives for supremacy
Under heaven
Intact,
His men and weapons
Still keen,
His gain
Complete.
This is the method of
Strategic attack.

In War,
With forces ten
To the enemy's one,
Surround him;
With five,
Attack him;
With two,
Split in half.
If equally matched,

Fight it out;
If fewer in number,
Lie low;
If weaker,
Escape.

A small force
    Obstinately fighting
    Will be captured
    By a larger force.

The general is the prop
    Of the nation.
When the prop is solid,
    The nation is strong.
When the prop is flawed,
    The nation is weak.

A ruler can bring misfortune
    Upon his troops
    In three ways:

    Ordering them
    To advance
    Or to retreat
    When they should not
    Is called
    Hobbling the army;

Ignorant interference
In military decisions
Confuses
Officers and men;

Ignorant meddling
In military appointments
Perplexes
Officers and men.

When an army is confused and perplexed,
The feudal princes
Will cause trouble;
        This creates
Chaos in the ranks
And gives away
Victory.

There are Five Essentials
For victory:

Know when to fight
And when not to fight;

Understand how to deploy
Large and small
Numbers;

Have officers and men who
Share a single will;

Be ready
For the unexpected;

Have a capable general,
Unhampered by his sovereign.

These five
Point the way to
Victory.

Hence the saying
'Know the enemy,
Know yourself,
And victory
Is never in doubt,
Not in a hundred battles.'

He who knows self
But not the enemy
Will suffer one defeat
For every victory.

He who knows
Neither self
Nor enemy
Will fail
In every battle.

Forms and Dispositions

Master Sun said:

Of old,
    The Skilful Warrior
    First ensured
    His own
    Invulnerability;
    Then he waited for
    The enemy's
    Vulnerability.

Invulnerability rests
    With self;
    Vulnerability,
    With the enemy.

The Skilful Warrior
    Can achieve
    His own
    Invulnerability;
    But he can never bring about
    The enemy's
    Vulnerability.

Hence the saying
    'One can know
    Victory
    And yet not achieve it.'

Invulnerability is
    Defence;
    Vulnerability is
    Attack.

Defence implies
    Lack;
    Attack implies
    Abundance.

A Skilful Defender
    Hides beneath
    The Ninefold Earth;
    A Skilful Attacker
    Moves above
    The Ninefold Heaven.

Thus they achieve
    Protection
And victory
    Intact.

To foresee
    The ordinary victory
    Of the common man
    Is no true skill.

To be victorious in battle
    And to be acclaimed
    For one's skill
    Is no true
    Skill.

To lift autumn fur
    Is no
    Strength;
To see sun and moon
    Is no
    Perception;
To hear thunder
    Is no
    Quickness of hearing.

The Skilful Warrior of old
    Won
    Easy victories.

The victories
    Of the Skilful Warrior
    Are not
    Extraordinary victories;
    They bring
    Neither fame for wisdom
    Nor merit for valour.

His victories
    Are
    Flawless;

His victory is
    Flawless
    Because it is
    Inevitable;
He vanquishes
    An already defeated enemy.

The Skilful Warrior
    Takes his stand
    On invulnerable ground;
    He lets slip no chance
    Of defeating the enemy.

The victorious army
    Is victorious first
    And seeks battle later;
    The defeated army
    Does battle first
    And seeks victory later.

The Skilful Strategist
    Cultivates
    The Way
    And preserves
    The law;
    Thus he is master
    Of victory and defeat.

In War,
> There are Five Steps:

> Measurement,
> Estimation,
> Calculation,
> Comparison,
> Victory.

Earth determines
> Measurement;
> Measurement determines
> Estimation;
> Estimation determines
> Calculation;
> Calculation determines
> Comparison;
> Comparison determines
> Victory.

A victorious army
> Is like a pound weight
> In the scale against
> A grain;
A defeated army
> Is like a grain
> In the scale against
> A pound weight.

A victorious army
    Is like
    Pent-up water
    Crashing
    A thousand fathoms
    Into a gorge.

This is all
    A matter of
    Forms and
    Dispositions.

Potential Energy

Master Sun said:

Managing many
    Is the same as
    Managing few;
    It is a question of
    Division.

Fighting with many
    Is the same as
    Fighting with few;
    It is a matter of
    Marshalling men
    With gongs,
    Identifying them
    With flags.

With a combination of
    Indirect and
    Direct,
    An army
    Can hold off the enemy
    Undefeated.

With an understanding of
    Weakness and
    Strength,
    An army
    Can strike
    Like a millstone
    Cast at an egg.

In warfare,
    Engage
    Directly;
    Secure victory
    Indirectly.

The warrior skilled
    In indirect warfare
    Is infinite
    As Heaven and Earth,
    Inexhaustible
    As river and sea,
    He ends and begins again
    Like sun and moon,
    Dies and is born again
    Like the Four Seasons.

There are but
    Five notes,
    And yet their permutations
    Are more
    Than can ever be heard.

There are but
    Five colours,
    And yet their permutations
    Are more
    Than can ever be seen.

There are but
    Five flavours,
    And yet their permutations
    Are more
    Than can ever be tasted.

In the dynamics of War,
    There are but these two—
    Indirect
    And direct—
    And yet their permutations
    Are inexhaustible.
    They give rise to each other
    In a never-ending,
    Inexhaustible circle.

A rushing torrent
    Carries boulders
    On its flood;
    Such is the energy
    Of its momentum.

A swooping falcon
    Breaks the back
    Of its prey;
    Such is the precision
    Of its timing.

The Skilful Warrior's energy is
    Devastating;
    His timing,
    Taut.

His energy is like
    A drawn crossbow,
    His timing like
    The release of a trigger.

In the tumult of battle,
    The struggle may seem
    Pell-mell,
    But there is no disorder;
    In the confusion of the melee,
    The battle array may seem
    Topsy-turvy,
    But defeat is out of the question.

Disorder is founded
    On order;
    Fear,
    On courage;
    Weakness,
    On strength.

Orderly disorder
    Is based on
    Careful division;
    Courageous fear,
    On potential energy;
    Strong weakness,
    On troop dispositions.

The warrior skilled at
    Stirring the enemy
    Provides a visible form,
    And the enemy is sure to come.
    He proffers the bait,
    And the enemy is sure
    To take it.
    He causes the enemy
    To make a move
    And awaits him
    With full force.

The Skilful Warrior
    Exploits
    The potential energy;
    He does not hold his men
    Responsible.
He deploys his men
    To their best
    But relies on
    The potential energy.

Relying on the energy,
    He sends his men into battle
    Like a man
    Rolling logs or boulders.
    By their nature,
    On level ground
    Logs and boulders
    Stay still;
    On steep ground
    They move;
    Square, they halt;
    Round, they roll.
    Skilfully deployed soldiers
    Are like round boulders
    Rolling down
    A mighty mountainside.

These are all matters
    Of potential energy.

Empty and Full

Master Sun said:

First on the battlefield
Waits for the enemy
Fresh.

Last on the battlefield
Charges into the fray
Exhausted.

The Skilful Warrior
Stirs
And is not stirred.

He lures his enemy
Into coming
Or obstructs him
From coming.

Exhaust
A fresh enemy;
Starve
A well-fed enemy;
Unsettle
A settled enemy.

Appear at the place
    To which he must hasten;
    Hasten to the place
    Where he least expects you.

March hundreds of miles
    Without tiring,
    By travelling
    Where no enemy is.

Be sure of victory
    By attacking
    The undefended.

Be sure of defence
    By defending
    The unattacked.

The Skilful Warrior attacks
    So that the enemy
    Cannot defend;
    He defends
    So that the enemy
    Cannot attack.

Oh, subtlety of subtleties!
    Without form!
Oh, mystery of mysteries!
    Without sound!
    He is master of
    His enemy's fate.

He advances
    Irresistibly,
    Attacking emptiness.

He retreats,
    Eluding pursuit,
    Too swift
    To be overtaken.

If I wish to engage,
    Then the enemy,
    For all his high ramparts
    And deep moat,
    Cannot avoid the engagement;
    I attack that which
    He is obliged
    To rescue.

If I do not wish to engage,
    I can hold my ground
    With nothing more than a line
    Drawn around it.
    The enemy cannot
    Engage me
    In combat:
    I distract him
    In a different direction.

His form is visible,
But I am
Formless;
I am concentrated,
He is divided.

I am concentrated
Into one;
He is divided
Into ten.
I am
Ten
To his one;
Many
Against
His few.

Attack few with many,
And my opponent
Will be weak.

The place I intend to attack
Must not be known;
If it is unknown,
The enemy will have to
Reinforce many places;
The enemy will
Reinforce many places,
But I shall attack
Few.

By reinforcing his vanguard,
    He weakens his rear;
By reinforcing his rear,
    He weakens his vanguard.
By reinforcing his right flank,
    He weakens his left;
By reinforcing his left,
    He weakens his right.
By reinforcing every part,
    He weakens every part.

Weakness
    Stems from
    Preparing against attack.
Strength
    Stems from
    Obliging the enemy
    To prepare against an attack.

If we know
    The place and the day
    Of the battle,
    Then we can engage
    Even after a march
    Of hundreds of miles.

But if neither day
    Nor place
    Is known,
    Then left cannot
    Help right,

Right cannot
Help left,
Vanguard cannot
Help rear,
Rear cannot
Help vanguard.
It is still worse
If the troops
Are separated
By a dozen miles
Or even by a mile or two.

According to my assessment,
The troops of Yue
Are many,
But that will avail them little
In the struggle.
So I say
Victory
Is still possible.

The enemy may be many,
But we can prevent
An engagement.

Scrutinize him,
Know the flaws
In his plans.

Rouse him,
    Discover the springs
    Of his actions.

Make his form visible,
    Discover his grounds
    Of death and life.

Probe him,
    Know his strengths
    And weaknesses.

The highest skill
    In forming dispositions
    Is to be without form;
    Formlessness
    Is proof against the prying
    Of the subtlest spy
    And the machinations
    Of the wisest brain.

Exploit the enemy's dispositions
    To attain victory;
    This the common man
    Cannot know.
    He understands
    The forms,
    The dispositions
    Of my victory;

But not
How I created the forms
Of victory.

Victorious campaigns
Are unrepeatable.
They take form in response
To the infinite varieties
Of circumstance.

Military dispositions
Take form like water.
Water shuns the high
And hastens to the low.
War shuns the strong
And attacks the weak.

Water shapes its current
From the lie of the land.
The warrior shapes his victory
From the dynamic of the enemy.

War has no
Constant dynamic;
Water has no
Constant form.

Supreme military skill lies
In deriving victory
From the changing circumstances
Of the enemy.

Among the Five Elements
    There is no one
    Constant supremacy.
The Four Seasons
    Have no
    Fixed station;
There are long days
    And short;
    The moon
    Waxes
    And it
    Wanes.

The Fray

Master Sun said:

In War,
 The general
 Receives orders
 From his sovereign,
 Assembles troops,
 And forms an army.
 He makes camp
 Opposite the enemy.
 The true difficulty
 Begins with
 The fray itself.

The difficulty of the fray
 Lies in making
 The crooked
 Straight
 And in making
 An advantage
 Of misfortune.

Take a roundabout route,
 And lure the enemy
 With some gain;

Set out after him,
    But arrive before him;
    This is to master
    The crooked
    And the straight.

The fray can bring
    Gain;
    It can bring
    Danger.

Throw your entire force
    Into the fray
    For some gain,
    And you may still
    Fail.

Abandon camp and
    Enter the fray
    For some gain,
    And you may lose
    Your equipment.

Order your men to
    Carry their armour
    And make forced march,
    Day and night,
    Without halting,
    March thirty miles
    At double speed
    For some gain,

And you will lose
All your commanders.
The most vigorous men
Will be in the vanguard;
The weakest,
In the rear.
One in ten
Will arrive.

March fifteen miles
For some gain,
And the commander
Of the vanguard
Will fall;
Only half the men
Will arrive.

March ten miles
For some gain,
And two in three men
Will arrive.

Without its equipment,
An army is lost;
Without provisions,
An army is lost;
Without base stores,
An army is lost.

Without knowing the plans
　　Of the feudal lords,
　　You cannot
　　Form alliances.

Without knowing the lie
　　Of hills and woods,
　　Of cliffs and crags,
　　Of marshes and fens,
　　You cannot
　　March

Without using local guides,
　　You cannot
　　Exploit
　　The lie of the land.

War
　　Is founded
　　On deception;
　　Movement is determined
　　By advantage;
　　Division and unity
　　Are its elements
　　Of Change.

Be rushing as a wind;
　　Be stately as a forest;
Be ravaging as a fire;
　　Be still as a mountain.

Be inscrutable as night;
    Be swift as thunder or lightning.

Plunder the countryside,
    And divide the spoil;
Extend territory,
    And distribute the profits.
Weigh the situation carefully
    Before making a move.

Victory belongs to the man
    Who can master
    The stratagem of
    The crooked
    And the straight.

This is the
    Art of the Fray.

The Military Primer says:

When ears do not hear,
    Use gongs and drums.
When eyes do not see,
    Use banners and flags.

Gongs and drums,
    Banners and flags
    Are the
    Ears and eyes
    Of the army.

With the army focused,
    The brave will not
    Advance alone,
    Nor will the fearful
    Retreat alone.

This is the Art of
    Managing Many.

In night fighting,
    Use torches and drums;
In daylight,
    Use banners and flags;
    So as to transform
    The ears and eyes
    Of the troops.

A whole fighting force
    Can be robbed
    Of its spirit;
A general
    Can be robbed
    Of his presence of mind.

The soldier's spirit
    Is keenest
    In the morning;
    By noon
    It has dulled;

By evening
He has begun
To think of home.

The Skilful Warrior
Avoids the keen spirit,
Attacks the dull
And the homesick;
This is
Mastery of Spirit.

He confronts chaos
With discipline;
He treats tumult
With calm.
This is
Mastery of Mind.

He meets distance
With closeness;
He meets exhaustion
With ease;
He meets hunger
With plenty;
This is
Mastery of Strength.

He does not intercept
Well-ordered banners;
He does not attack
A perfect formation.

This is
　　Mastery of Change.

These are axioms
　　Of the Art of War:
　　Do not advance uphill.
　　Do not oppose an enemy
　　With his back to a hill.
　　Do not pursue an enemy
　　Feigning flight.
　　Do not attack
　　Keen troops.
　　Do not swallow
　　A bait.
　　Do not thwart
　　A returning army.

　　Leave a passage
　　For a besieged army.

　　Do not press
　　An enemy at bay.

This is
　　The Art of War.

The Nine Changes

Master Sun said:

In War,
    The general
    Receives orders
    From his sovereign,
    Then assembles troops
    And forms an army.

On intractable terrain,
    Do not encamp;
On crossroad terrain,
    Join forces with allies;
On dire terrain,
    Do not linger;
On enclosed terrain,
    Make strategic plans;
On death terrain,
    Do battle.

There are roads
    Not to take.
There are armies
    Not to attack.

There are towns
> Not to besiege.
There are terrains
> Not to contest.
There are ruler's orders
> Not to obey.

The general
> Who knows the gains
> Of the Nine Changes
> Understands War.

The general
> Ignorant of the gains
> Of the Nine Changes
> May know the lie of the land,
> But he will never reap
> The gain
> Of that knowledge.

The warrior
> Ignorant of the Art
> Of the Nine Changes,
> May know
> The Five Gains
> But will not get the most
> From his men.

The wise leader
    In his deliberations
    Always blends consideration
    Of gain
    And harm.

By tempering thoughts of
    Gain,
    He can accomplish
    His goal;

By tempering thoughts of
    Harm,
    He can extricate himself
    From calamity.

He reduces the feudal lords
    To submission
    By causing them
    Harm;

He wears them down
    By keeping them
    Constantly occupied;

He precipitates them
    With thoughts of
    Gain.

The Skilful Warrior
     Does not rely on the enemy's
     Not coming,
     But on his own
     Preparedness.

     He does not rely on the enemy's
     Not attacking,
     But on his own
     Impregnability.

There are Five Pitfalls
     For a general:

Recklessness,
     Leading to
     Destruction;
Cowardice,
     Leading to
     Capture;
A hot temper,
     Prone to
     Provocation;
A delicacy of honour,
     Tending to
     Shame;
A concern for his men,
     Leading to
     Trouble.

These Five Excesses
    In a general
    Are the
    Bane of war.

If an army is defeated
    And its general slain,
    It will surely be because of
    These Five Perils.
They demand the most
    Careful consideration.

On the March

Master Sun said:

In taking up position
    And confronting the enemy:

Cross mountains,
    Stay close to valleys;
    Camp high,
    And face the open;
    Fight downhill,
    Not up.
These are positions in
    Mountain warfare.

Cross rivers,
    Then keep a distance
    From them.
    If the enemy crosses a river
    Towards you,
    Do not confront him
    In midstream.
    Let half his troops cross
    Before you strike.
    If you wish to do battle,
    Do not confront the enemy

Close to the river.
Occupy high ground,
And face the open.
Do not advance
Against the flow.
These are positions in
River warfare.

Cross salt marshes
Rapidly;
Never linger.
If you must do battle
In a salt marsh,
Keep water plants
Close by
And trees
Behind you.
These are positions in
Salt marshes.

On level ground,
Occupy easy terrain.
Keep high land
To the right and rear:
Keep death in front
And life to the rear.
These are positions on
Level ground.

Observation of
    These four types of positions
    Enabled the Yellow Emperor
    To defeat
    The Four Emperors.

Armies prize high ground,
    Shun low;
    They esteem Yang,
    Avoid Yin.

Nurture life,
    Occupy solid ground.
    Your troops will thrive,
    Victory will be sure.

On mound,
    Hill,
    Bank,
    Or dike,
    Occupy the Yang,
    With high ground
    To right and rear.
Use the lie of the land
    To the troops' benefit.

When rains upstream
    Have swollen the river,
    Let the water subside
    Before crossing.

If you come to
> Heaven's Torrents,
> Heaven's Wells,
> Heaven's Prisons,
> Heaven's Nets,
> Heaven's Traps,
> Heaven's Cracks:
> Quit such places
> With all speed.
> Do not go near them.
> Keep well away,
> Let the enemy
> Go near them.
> Keep them in front;
> Let him have them
> At his rear.

If you march by
> Ravine,
> Swamp,
> Reedy marshland,
> Mountain forest,
> Thick undergrowth:
> Beware,
> Explore them diligently.
> These are places
> Of ambush,
> Lairs for spies.

When the enemy is
> Close at hand
> And makes no move,
> He is counting on
> A strong position;

If he is
> At a distance
> And provokes battle,
> He wants his opponent
> To advance.

If he is
> On easy ground,
> He is luring us.

If trees move,
> He is coming.

If there are many screens
> In the grass,
> He wants
> To perplex us.

Birds rising in flight
> Are a sign
> Of ambush;

Beasts startled
> Are a sign
> Of surprise attack.

Dust high and peaking
    Is a sign
    Of chariots approaching;

Dust low and spreading
    Is a sign
    Of infantry approaching;

Dust in scattered strands
    Is a sign
    Of firewood's being collected;

Dust in drifting pockets
    Is a sign
    Of an army encamping.

Humble words, coupled with
    Increased preparations,
    Are a sign
    Of impending attack;

Strong words, coupled with
    An aggressive advance,
    Are a sign
    Of impending retreat.

Light chariots
    Emerging first
    On the wings
    Are a sign
    Of battle formation.

Words of peace,
    But no treaty,
    Are a sign
    Of a plot.

Much running about
    And soldiers parading
    Are a sign
    Of expectation.

Some men advancing
    And some retreating
    Are a sign
    Of a decoy.

Soldiers standing
    Bent on their spears
    Indicate great
    Hunger.

Bearers of water
    Drinking first
    Indicate great
    Thirst.

An advantage perceived,
    But not acted on,
    Indicates utter
    Exhaustion.

Birds gather
　　On empty ground.

Shouting at night
　　Is a sign
　　Of fear.

Confusion among troops
　　Is a sign
　　That the general
　　Is not respected.

Banners and flags moving
　　Are a sign
　　Of disorder.

If officers
　　Are prone to anger,
　　The men become weary.

If they feed
　　Grain to their horses
　　And meat to their men;
If they fail to
　　Hang up their pots
　　And do not
　　Return to their quarters;
　　Then they are
　　At bay.

Men whispering together,
    Huddled in small groups,
    Are a sign
    Of disaffection.

Excessive rewards
    Are a sign
    Of desperation.

Excessive punishments
    Are a sign
    Of exhaustion.

If a general is by turns
    Tyrannical
    And in terror
    Of his own men,
    It is a sign of
    Supreme incompetence.

Envoys
    With words of conciliation
    Desire cessation.

Protracted, fierce
    Confrontation,
    With neither engagement
    Nor retreat,
    Must be regarded
    With great vigilance.

In War,
 Numbers
 Are not the issue.
It is a question of
 Not attacking
 Too aggressively.
 Concentrate your strength,
 Assess your enemy,
 And win the confidence of your men:
 That is enough.

Rashly underestimate your enemy,
 And you will surely be
 Taken captive.

Discipline troops
 Before they are loyal,
 And they will be
 Refractory
 And hard to put to good use.
 Let loyal troops
 Go undisciplined,
 And they will be altogether
 Useless.

Command them
 With civility,
 Rally them
 With martial discipline,
 And you will win their
 Confidence.

Consistent and effective orders
    Inspire obedience;
    Inconsistent and ineffective orders
    Provoke disobedience.

When orders are consistent
    And effective,
    General and troops
    Enjoy mutual trust.

Forms of Terrain

Master Sun said:

There are different forms of terrain:

> Accessible terrain,
> Entangling terrain,
> Deadlock terrain,
> Enclosed terrain,
> Precipitous terrain,
> Distant terrain.

'Accessible' means that
> Both sides
> Can come and go freely.
> On accessible terrain,
> He who occupies
> High Yang ground
> And ensures
> His line of supplies
> Will fight
> To advantage.

'Entangling' means that
> Advance is possible,
> Withdrawal hard.

On entangling terrain,
If the enemy is unprepared,
Go out and defeat him.
But if he is prepared,
And our move fails,
It will be hard to retreat.
The outcome will not be
To our advantage.

'Deadlock' means that
Neither side finds it
Advantageous
To make a move.
On deadlock terrain,
Even if our enemy
Offers a bait,
We do not make a move;
We lure him out;
We retreat.
And when half his troops
Are out,
That is our moment
To strike.

On enclosed terrain,
If we occupy it first,
We must block it
And wait for the enemy.
If he occupies it first
And blocks it,
Do not go after him;

If he does not block it,
Then go after him.

On precipitous terrain,
    If we occupy it first,
    We should hold the Yang heights
    And wait for the enemy.
    If the enemy occupies it first,
    Do not go after him,
    But entice him out
    By retreating.

On distant terrain,
    When strengths are matched,
    It is hard to provoke battle,
    And an engagement
    Will not be advantageous.

These six constitute
    The Way of Terrain.
    It is the general's duty
    To study them diligently.

In War,
    The following are not
    Natural calamities,
    But the fault
    Of the general:

Flight
Impotence,
Decay,
Collapse,
Chaos,
Rout.

If relative strengths are matched,
But one army faces another
Ten times its size,
The outcome is
Flight.

When troops are strong
But officers weak,
The result is
Impotence.

When officers are strong
But troops weak,
The result is
Decay.

When superior officers are angry
And insubordinate
And charge into battle
Out of resentment,
Before their general can judge
The likelihood of victory,
Then the outcome is
Collapse.

When the general is weak
    And lacking in severity,
    When his orders
    Are not clear,
    When neither officers nor men
    Have fixed rules
    And troops
    Are slovenly,
    The outcome is
    Chaos.

When a general
    Misjudges his enemy
    And sends a lesser force
    Against a larger one,
    A weaker contingent
    Against a stronger one;
    When he fails to pick
    A good vanguard,
    The outcome is
    Rout.

These six constitute
    The Way of Defeat.
    It is the general's duty
    To study them diligently.

The form of the terrain
    Is the soldier's ally;

Assessment
    Of the enemy
    And mastery of victory;
    Calculating the difficulty,
    The danger,
    And the distance
    Of the terrain;
    These constitute the Way
    Of the Superior General.

He who knows this
    And practises it in battle
    Will surely be
    Victorious.
    He who does not know it
    And does not practise it
    Will surely be
    Defeated.

If an engagement is sure
    To bring victory,
    And yet the ruler
    Forbids it,
    Fight;
    If an engagement is sure
    To bring defeat,
    And yet the ruler
    Orders it,
    Do not fight.

He who advances
    Without seeking
    Fame,
    Who retreats
    Without escaping
    Blame,
    He whose one aim is
    To protect his people
    And serve his lord,
    This man is
    A Jewel of the Realm.

He regards his troops
    As his children,
    And they will go with him
    Into the deepest ravine.
    He regards them
    As his loved ones,
    And they will stand by him
    Unto death.

If he is generous
    But cannot command,
If he is affectionate
    But cannot give orders,
If he is chaotic
    And cannot keep order,
Then his men
    Will be like
    Spoiled children,
    And useless.

If we know that our own troops
    Are capable of attacking
    But fail to see
    That the enemy
    Is not vulnerable,
    We have only
    Half of victory.

If we know that the enemy
    Is vulnerable
    But fail to see
    That our own troops
    Are incapable of attacking,
    We have only
    Half of victory.

If we know that the enemy
    Is vulnerable,
    And know that our own troops
    Are capable of attacking,
    But fail to see
    That the terrain
    Is unfit for attack,
    We still have only
    Half of victory.

The Wise Warrior,
    When he moves,
    Is never confused;
    When he acts,
    Is never at a loss.

So it is said;
    'Know the enemy,
    Know yourself,
    And victory
    Is never in doubt,
    Not in a hundred battles.'

    Know Heaven,
    Know Earth,
    And your victory
    Is complete.

The Nine Kinds of Ground

Master Sun said:

In War,
   There are
   Nine Kinds of Ground:

   Scattering ground,
   Light ground,
   Strategic ground,
   Open ground,
   Crossroad ground,
   Heavy ground,
   Intractable ground,
   Enclosed ground,
   Death ground.

When the feudal lords
   Fight on home territory,
   That is
   Scattering ground.

When an army enters
   Enemy territory,
   But not deeply,

That is
Light ground.

When the ground
    Offers advantage
    To either side,
    That is
    Strategic ground.

When each side
    Can come and go freely,
    That is
    Open ground.

When the ground
    Borders
    Three states
    And the first to take it
    Has mastery
    Of the empire,
    That is
    Crossroad ground.

When an army enters
    Enemy territory deeply
    And holds
    Several fortified towns
    In its rear,
    That is
    Heavy ground.

When an army travels through
    Mountains and forests,
    Cliffs and crags,
    Marshes and fens,
    Hard roads,
    These are
    Intractable ground.

Ground reached
    Through narrow gorges,
    Retreated from
    By twisting paths,
    Where a smaller force of theirs
    Can strike our larger one,
    That is
    Enclosed ground.

Ground where mere survival
    Requires
    A desperate struggle,
    Where without
    A desperate struggle
    We perish,
    That is
    Death ground.

On scattering ground,
    Do not fight.
On light ground,
    Do not halt.

On strategic ground,
    Do not attack.
On open ground,
    Do not block.
On crossroad ground,
    Form alliances.
On heavy ground,
    Plunder.
On intractable ground,
    Keep marching.
On enclosed ground,
    Devise stratagems.
On death ground,
    Fight.

The Skilful Warrior of old
    Could prevent
    The enemy's vanguard
    From linking with his rear,
    Large and small divisions
    From working together,
    Crack troops
    From helping poor troops,
    Officers and men
    From supporting one another.
    The enemy,
    Once separated,
    Could not
    Reassemble;

Once united,
Could not
Act in concert.

When there was some gain
To be had,
He made a move;
When there was none,
He halted.

To the question
'How should we confront
Numerous and well arrayed,
Poised to attack?'
My reply is
'Seize something
He cherishes,
And he will do your will.'

Speed
Is the essence of War.
Exploit the enemy's unpreparedness;
Attack him unawares;
Take an unexpected route.

The Way of Invasion is this:
Deep penetration
Brings cohesion;
Your enemy
Will not prevail.

Plunder fertile country
    To nourish your men.
    Cherish your troops,
    Do not wear them out.
    Nurture your energy;
    Concentrate it.

Move your men about;
    Devise stratagems
    That cannot be fathomed.
    Throw your men
    Where there is no escape,
    And they will die
    Rather than flee.
    Men who have
    Faced death
    Can achieve anything;
    They will give
    Their last drop of strength,
    Officers and men alike.

Troops in desperate straits
    Know no fear.
    Where there is no escape,
    They stand firm;
    When they have entered deep,
    They persist;
    When they see no hope,
    They fight.

They are alert
    Without needing
    Discipline;
    They act
    Without needing
    Instructions;
    They are devoted
    Without needing
    A compact;
    They are loyal
    Without needing
    Orders.

Forbid the consulting of omens,
    Cast out doubts,
    And they will go on
    To the death.

Our men have no excess
    Of worldly goods,
    And yet they do not
    Disdain wealth;
    They do not expect
    To live long,
    And yet they do not
    Disdain long life.

On the day
    They are ordered into battle,
    They sit up and weep,
    Wetting their clothes with their tears;

They lie down and weep,
Wetting their cheeks.

But throw them
    Where there is no escape,
    And they will fight
    With the courage
    Of the heroes
    Zhu and Gui.

The Skilful Warrior
    Deploys his troops
    Like the *shuairan* snake
    Found on Mount Heng.
    Strike its head,
    And the tail lashes back;
    Strike its tail,
    And the head fights back;
    Strike its belly,
    And both head and tail
    Will attack you.
    To the question
    'Can an army be
    Like the *shuairan* snake?'
    I reply,
    'Yes, it can.'
    Take the men of Wu
    And the men of Yue.
    They are enemies,
    But if they cross a river
    In the same boat

And encounter a wind,
They will help each other,
Like right hand and left.

It is not enough
     To tether horses
     And to bury
     Chariot wheels.

There must be a single courage
     Throughout:
     This is the Way
     To manage an Army.

Strong and weak,
     Both can serve,
     Thanks to the principle
     Of ground.

The Skilful Warrior
     Directs his army
     As if it were
     A single man.
     He leaves it no choice
     But to obey.

It is the business of the general
     To be still
     And inscrutable,
     To be upright
     And impartial.

He must be able
    To keep his own troops
    In ignorance,
    To deceive their eyes
    And their ears.

He changes his ways
    And alters his plans
    To keep the enemy
    In ignorance.

He shifts camp
    And takes roundabout routes
    To keep the enemy
    In the dark.

He leads his men into battle
    Like a man
    Climbing a height
    And kicking away the ladder;
    He leads them
    Deep into the territory
    Of the feudal lords
    And releases the trigger.
    He burns his boats,
    He breaks his pots.
    He is like a shepherd
    Driving his sheep
    This way and that;
    No one knows
    Where he is going.

He assembles his troops
    And throws them
    Into danger;
    This is the business
    Of the commander.

These things must be studied:
    The Variations
    Of the Nine Kinds of Ground;
    The Advantages
    Of Flexible Manouvre;
    The Principles
    Of Human Nature.

The Way of Invasion is this:
    Deep penetration
    Brings cohesion;
    Shallow penetration
    Brings scattering.

When you leave your own territory
    And lead your men
    Across the border,
    You enter dire terrain.

When there are lines of communication
    On all four sides,
    You are on
    Crossroad terrain.

When you penetrate deeply,
     You are on
     Heavy terrain.

When you penetrate superficially,
     You are on
     Light terrain.

When there are strongholds to your rear
     And narrow passes in front,
     You are on
     Enclosed terrain.

When there is no way out,
     You are on
     Death terrain.

On scattering ground,
     We unite the will of our men.

On light ground,
     We keep them connected.

On strategic ground,
     We bring up our rear.

On open ground,
     We see to our defences.

On crossroad ground,
     We strengthen our alliances.

On heavy ground,
    We ensure continuity of supplies.

On intractable ground,
    We keep on the move.

On enclosed ground,
    We block the passes.

On death ground,
    We demonstrate
    The desperateness
    Of the situation.

It is in the soldier's nature that
    When surrounded,
    He resists;
    When all seems lost,
    He struggles on;
    When in danger,
    He obeys orders.

Without knowing the plans
    Of the feudal lords,
    You cannot
    Form alliances.

Without knowing the lie
    Of hills and woods,
    Of cliffs and crags,
    Of marshes and fens,

You cannot
    March.

Without using local guides,
    You cannot
    Exploit
    The lie of the land.

Ignorance of any one
    Of these points
    Is not characteristic
    Of the army of a great king.

When the army of a great king
    Attacks a powerful state,
    He does not allow the enemy
    To concentrate his forces.
    He overawes the enemy
    And undermines his alliances.

He does not strive
    To ally himself
    With all the other states;
He does not foster
    Their power;
He pursues
    His own secret designs,
    Overawing his enemies.

Thus he can capture
    The enemy's cities
    And destroy
    The enemy's state.

Distribute rewards
    Without undue respect for rules;
    Publish orders
    Without undue regard for precedent;

Deal with a whole army
    As if it were a single man.
    Apply them to their task
    Without words of explanation.
    Confront them with the advantage,
    But do not explain the danger.

Throw them into
    Perilous ground,
    And they will survive;
    Plunge them into
    Death ground,
    And they will live.

When a force
    Has fallen into danger,
    It can
    Snatch victory
    From defeat.

Success in war
Lies in
Scrutinizing
Enemy intentions.
And going with them.

Focus on the enemy,
And from hundreds of miles
You can kill their general.
This is
Success
Through cunning.

On the day
You decide to attack,
Close the passes,
Destroy the tallies,
Break off intercourse
With envoys;
Be firm in the temple council
For the execution of
Your plans.

If the enemy opens a door,
Rush in.
Seize what he holds dear,
And secretly contrive
An encounter.

Discard rules,
    Follow the enemy,
    To fight
    The decisive battle.

At first,
    Be like a maiden;
    When the enemy opens the door,
    Be swift as a hare;
    Your enemy will not
    Withstand you.

Attack by Fire

Master Sun said:

> There are Five Ways to
> Attack by Fire.
>
> The first is to burn
> Men;
>
> The second is to burn
> Supplies;
>
> The third is to burn
> Equipment;
>
> The fourth is to burn
> Warehouses;
>
> The fifth is to burn
> Lines of communication.

Attack by fire
> Requires means;
> The material
> Must be ready.

There is a season
    For making a fire;
    There are days
    For lighting a flame.

The proper season is
    When the weather is
    Hot and dry;

The proper days are
    When the moon is in
    Sagittarius,
    Pegasus,
    Crater,
    Corvus.
    These are the
    Four Constellations
    Of Rising Wind.

When attacking with fire,
    Adapt to
    These Five Changes of Fire:

If fire breaks out
    Within the enemy camp,
    Respond at once
    From without.

If fire breaks out
    But the enemy remains calm,
    Wait,
    Do not attack.
    Let the fire reach
    Its height,
    And follow up
    If at all possible;
    If not,
    Wait.

If fire attack is possible
    From without,
    Do not wait
    For fire to be started
    Within;
    Light
    When the time is right.

When starting a fire,
    Be upwind;
    Never attack
    From downwind.

A wind that rises
    During the day
    Lasts long;
    A night wind
    Soon fails.

In War,
    Know these
    Five Changes of Fire,
    And be vigilant.

Fire
    Assists an attack
    Mightily.

Water
    Assists an attack
    Powerfully.

Water
    Can isolate,
    But it cannot
    Take away.

To win victory,
    To complete an objective,
    But not to follow through,
    Is a disastrous
    Waste.

Hence the saying
    'The enlightened ruler
    Considers deeply;
    The effective general
    Follows through.'

Never move
    Except for gain;
Never deploy
    Except for victory;
Never fight
    Except in a crisis.

A ruler
    Must never
    Mobilize his men
    Out of anger;
    A general
    Must never
    Engage battle
    Out of spite.

Move
    If there is gain;
Halt
    If there is no gain.

Anger
    Can turn to
    Pleasure;
Spite
    Can turn to
    Joy.
But a nation destroyed
    Cannot be
    Put back together again;

A dead man
Cannot be
Brought back to life.

So the enlightened ruler
    Is prudent;
The effective general
    Is cautious.
This is the Way
    To keep a nation
    At peace
    And an army
    Intact.

Espionage

Master Sun said:

> Raising an army
> Of a hundred thousand men
> And marching them
> Three hundred miles
> Drains the pockets
> Of the common people
> And the public treasury
> To the daily sum of
> A thousand taels of silver.
> It causes commotion
> At home and abroad
> And sets countless men
> Tramping the highways
> Exhausted.
> It keeps seven hundred thousand families
> From their work.

> Two armies may
> Confront each other
> For several years,
> For a single
> Decisive battle.

It is callous
>   To begrudge the expense of
>   A hundred taels
>   Of silver
>   For knowledge
>   Of the enemy's situation.

Such a miser is
>   No commander of men,
>   No support to his lord,
>   No master of victory.

Prior information
>   Enables wise rulers
>   And worthy generals
>   To move
>   And conquer,
>   Brings them success
>   Beyond that of the multitude.

This information
>   Cannot be obtained
>   From spirits;
>   It cannot be deduced
>   By analogy;
>   It cannot be calculated
>   By measurement.

It can be obtained only
>From men,
>From those who know
The enemy's dispositions.

There are Five Sorts of Spies:

Local,
Internal,
Double,
Dead, and
Live.

When these five sorts of espionage
Are in operation,
No one knows
The Way of it.
It is called
The Mysterious Skein,
The Lord's Treasure.

Local spies
Come from among our enemy's
Fellow countrymen;

Internal spies,
>From among our enemy's
Officials,

Double spies,
    From among our enemy's
    Own spies.

Dead spies
    Are those for whom
    We deliberately create
    False information;
    They then pass it on
    To the enemy.

Live spies
    Are those who return
    With information.

In the whole army,
    None should be closer
    To the commander
    Than his spies,
    None more highly rewarded,
    None more confidentially treated.

Without wisdom,
    It is impossible
    To employ spies.

Without humanity and justice,
    It is impossible
    To employ spies.

Without subtlety and ingenuity,
    It is impossible
    To ascertain
    The truth of their reports.

Subtlety of subtleties!
    Spies have
    Innumerable uses.

If confidential information
    Is prematurely divulged,
    Both spy and recipient
    Must be put to death.

In striking an army,
    In attacking a city,
    In killing an individual,
    It is necessary to know beforehand
    The names of the general
    And of his attendants,
    His aides,
    His doorkeepers,
    His bodyguards.
    Our spies must be instructed
    To discover all of these
    In detail.

Enemy spies,
    Come to spy on us,
    Must be sought out,
    Bribed,

Won over,
Well accommodated.
Then they can be
Employed as
Double agents.

From the double agent
We discover
Local and internal spies.

From the double agent
We learn how best
To convey misinformation
To the enemy.

From the double agent
We know how and when
To use
Live spies.

The ruler
Must know all five of these
Sorts of spies;
This knowledge must come
From the double agent;
So the double agent
Must be
Treated generously.

Of old,
    The rise of the Yin dynasty
    Was due to Yi Zhi,
    Who had served under the Xia;
    And the rise of the Zhou dynasty
    Was due to Lü Ya,
    Who had served under the Yin.

Only the enlightened ruler,
    The worthy general,
    Can use
    The highest intelligence
    For spying,
    Thereby achieving
    Great success.

Spies
    Are a key element
    In warfare.
    On them depends
    An army's
    Every move.